Disney's Numbers Are Fun

by
Jane Werner Watson

Consultant:
Harold R. Melnick
Advisor in Graduate Programs and Instructor
Bank Street College of Education

® GOLDEN PRESS ● NEW YORK
Western Publishing Company, Inc.
Racine, Wisconsin

Printed in the U.S.A. by Western Publishing Company, Inc.
Golden, A Golden Book® and Golden Press® are trademarks of
Western Publishing Company, Inc.
Library of Congress Catalog Card Number: 77-71148

NUMBERS ARE FUN.

First comes 1.

Mickey Mouse wants to go on a picnic.
But Mickey is the only one here.
A picnic for one is not much fun.

Second comes **2**
for me and you.

Here comes Minnie Mouse.
Mickey and Minnie make two.

Third comes 3.
Yes sirree!

Mickey and Minnie and Morty Mouse
make one, two, three for the picnic.

Then one more gives us **4.**

Mickey and Minnie
and Morty and Ferdie
make four—one, two, three, four.

Man alive! Here come 5!

Donald and Daisy Duck

and Huey and Dewey
and Louie Duck make five—
one, two, three, four, five.

Quick! Quick! Climb on, 6!

Daisy and Donald Duck
and Mickey and Minnie
and Morty and Ferdie Mouse
make six.

Merciful heaven!
Is there room for 7?

Mickey and Minnie and Morty and Ferdie Mouse
and Donald and Daisy and Huey Duck
make seven.

Wait! Wait! We'll fit in 8!

Here comes Louie Duck.
He wants to go on the picnic, too.
Seven and one make eight.

Just in time, here comes 9.

Mickey and Minnie and Morty and Ferdie,
Donald and Daisy,
and Huey and Dewey and Louie make nine.
Nine on a picnic can have a fine time.

At the end,
Pluto pup makes 10.

Four Mice, five Ducks,
plus good old Pluto Pup—
One, two, three, four,
five, six, seven, eight, nine and one make ten,
all on their way to a picnic again.

SNOW WHITE'S FRIENDS

Snow White was alone
in the big forest.
She was afraid.
But forest friends
came out to help her.

How many can you see?

They led her to the home
of the seven dwarfs.

The seven dwarfs
(count them)
are happy to have
Snow White in their home.
They are having a party
to celebrate.
How many dwarfs are
playing music?
How many are dancing?
Are more dwarfs dancing
or playing music?
How many deer are watching?
How many birds and rabbits
and squirrels are watching?
Are there more animals
or dwarfs?

PINOCCHIO FISHING

Pinocchio
is busy fishing.
How many
fish can he catch?

Geppetto says,
"Ask some friends
for dinner.
I will cook
the fish.
Each person
will have a fish."
How many friends
can Pinocchio ask?
(Figaro, the kitten,
will be very happy
with the fish heads.)

CINDERELLA, DRESSMAKER

Cinderella's best friends
are the little mice
who live in her house.
She likes to make
little dresses and suits
for them.
"You must all have
new clothes for the ball,"
she says.

See how many little dresses
and suits she has finished.
How many more must she make?

THE PIRATES OF NEVER LAND

Captain Hook and his pirates are creeping through the woods of Never Land

to capture Wendy and Peter and John and Michael and the Lost Boys.

If each pirate can carry one child, how many pirates are needed?

Do you think there are enough pirates to carry all the children?

How many pirates can you see? How many children?

LADY AND THE TRAMP

Lady has come to live
in a new home.
She does not know anyone
in the neighborhood.
"Maybe I am the one and
only dog on the street,"
she says to herself.

But that is not so.
In the house next door
lives a Scottie
known as Jock.
"I am glad to know you,"
says Lady.
"One and one make_____.

$$1 + 1 = 2$$

In the house
on the other side
lives Trusty, the bloodhound.
"Oh, I see!" barks Lady.
"Two and one make_____." $2 + 1 = 3$

A carriage rolls
along the street.
Behind it runs
a proud carriage dog.
"Here's one more,"
says Lady.
"Three and one make_____."
$3 + 1 = 4$

Carriage dogs like
to run in pairs.
So along comes another.
"Well," says Lady, "sakes alive!
Four and one make_____." $4 + 1 = 5$

At the heels of the carriage dogs
trots a cocky mongrel.
His name is Tramp.
He stops to show
Lady his tricks.
Five and one, you know,
make_____. $5 + 1 = 6$

It is Tramp who,
in the end,
becomes Lady's
special friend.

MOWGLI'S NEW HOME

Mowgli, the jungle boy,
has lived with a wolf family.
Tonight the wolf pack
is meeting under the moon.
How many wolves can you count?

The animal that is not a wolf
is Bagheera, the black panther.
The wolves have decided
it is time for Mowgli
to live with his own kind.
Bagheera agrees to take Mowgli
to the man-village
just outside the jungle.
Off they go, one, two.

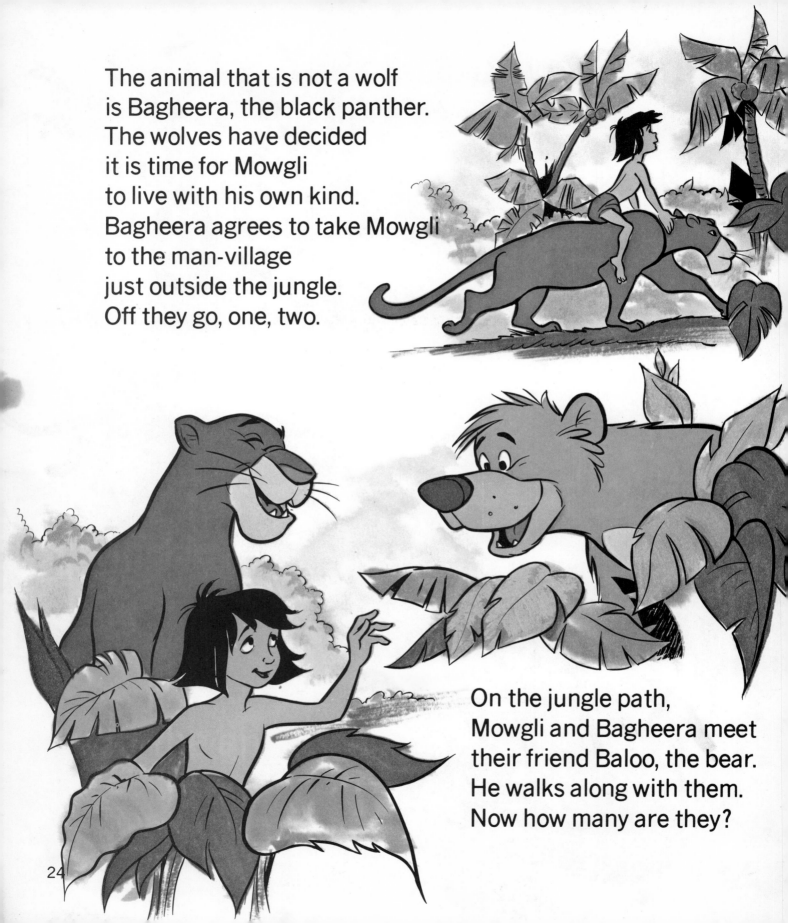

On the jungle path,
Mowgli and Bagheera meet
their friend Baloo, the bear.
He walks along with them.
Now how many are they?

Can you see some enemies
peeking through the leaves?
One is Kaa, the python snake.
One is Shere Khan, the tiger.
One and one make two, you know.
They will not harm Mowgli, though,
while he is with Bagheera and Baloo.

Mowgli and Baloo
stop to swim
in a jungle pool.
How many fish can you see
in the pool?
How many monkeys
can you see
in the trees?
There are birds
in the trees, too.
How many birds?

The monkeys snatch Mowgli!
They carry him off
to the monkey king
in the ruined city.
The monkey king
is eating bananas.
Can you tell
how many bananas he has eaten?
How many does he have left to eat?

Bagheera and Baloo
rescue Mowgli from the monkeys.
They take him to the edge
of the jungle.
There they can see
the man-village
where Mowgli will make
his new home.
How many houses
can you see?
Which one do you think
will be Mowgli's home?

BAMBI'S FRIENDS

Bambi, the baby deer,
is the Prince of the Forest.
His best friends are
Thumper, the rabbit,
Flower, the skunk,
and Faline, another little deer.
Here they all are—
one, two, three, four.

1   2   3   4

Winter comes to the forest.
Flower goes off
for a long winter's sleep.
How many are left
to play?

1   2   3

Faline's mother calls her.
So off she goes.
How many are left now?

1   2

Thumper goes off
to hunt for food.
How many are left now?

1

29

TEN LITTLE MONKEYS

Ten little monkeys
Swung on jungle vines.

One lost his grip,
And then there were nine.

$$10 - 1 = 9$$

Nine little monkeys
Marching through a gate.

The gate swung shut on one,
And then there were eight.

$$9 - 1 = 8$$

Eight little monkeys
Climbing up toward heaven.

One tumbled down,
And then there were seven. $8 - 1 = 7$

Seven little monkeys
Doing monkey tricks.

One fell upon his nose,
And then there were six. $7 - 1 = 6$

Six little monkeys
Saw a beehive.

One went too close to it,
And then there were five. $6 - 1 = 5$

Five little monkeys
Started to snore.

One couldn't stand the noise,
And then there were four. $5 - 1 = 4$

Four little monkeys
Eating hungrily.

One had enough at last,
And then there were three. $4 - 1 = 3$

Three little monkeys
Heard a call, "Yoo hoo!"

One went to answer it,
And then there were two. $\boxed{3 - 1 = 2}$

Two little monkeys
Playing in the sun.

Since the game was hide-and-seek,
Soon there was only one. $\boxed{2 - 1 = 1}$

One little monkey
Grew tired of being alone.

He went to find the others,
And then there were none. $1 - 1 = 0$

THREE LITTLE PIGS

This is the house
of Mrs. Pig.
It is a nice house.
But it is not big.

Hear all that noise?
"Toot!"
"Bang!"
"Scree scree!"
That's Mrs. Pig's sons,
One, Two and Three.

Mrs. Pig says,
"One, Two, and Three,
I must live
more quietly!"

So down the road
the little pigs roam,
each to build
himself a home.

Pig One builds
his house of straw,
the fastest built house
you ever saw.

Mrs. Pig's house
and the house of hay—
how many houses
can you count today?

42

Little Pig Two
builds a house of twigs.
Then he dances
some lively jigs.

Mrs. Pig's house,
one of straw,
one of sticks.
How many houses?
Count them quick!

Little Pig Three
builds his house of bricks.
It's hard, slow work,
without any tricks.

It's finished at last.
It is not very big.

One of brick,
one of straw,
one of sticks—
and Mrs. Pig's.
How many houses
now do you see?
Be sure you count
them carefully!

One, two, three
and one house more.
Yes, you're right.
There are really four.

Here comes the wolf!
With a huff and puff,
he turns the little
straw house to fluff!

How many houses
now do you see?
There are four no more,
just one, two, _____.

Away runs Pig One
to the house of his brother.

Wolf follows him,
and puffs at another.
Down goes the neat
little house of twigs.

Away run two
little frightened pigs!

Count the houses
now, can you?
Three minus one
make just one, _____ .

Here comes Wolf
to the house of bricks.
He huffs and he puffs
and tries all kinds of tricks.
He puffs and he huffs
and he puffs some more.
But he can't shake the house
or open the door.

Away slinks Wolf.
Out come the pigs.
They toot and they fiddle
and dance some jigs.

Now they live in the houses
you can see—
one for Mrs. Pig,
one for piglets three.

Who would wish
for anything more
than _____ little houses
for happy pigs _____?

NUMBERS APLENTY

1, 2
One, two,
Buckle my shoe.

Prince Charming finds that the glass slipper fits Cinderella.

3, 4
Three, four,
Shut the door.

Peter and his friends Sasha, the bird, and Sonia, the duck, shut the door on the wicked wolf of the wild forest.

5, 6
Five, six,
Pick up sticks.

Five grown-up elephants and one baby
march through the jungle in single file.

49

7, 8
Seven, eight,
Write on a slate.

TODAY I AM A REAL BOY

Geppetto, the woodcarver, and his friends Figaro, the kitten,
and Cleo, the goldfish, are so pleased! Pinocchio, the puppet,
has earned the right to be a real, live boy! Jiminy Cricket
is pleased, too. So is the Blue Fairy, who made it happen.
Everyone is pleased except bad old Foulfellow, the fox,
and Gideon, the cat.

9, 10
Nine, ten,
Big fat hens.

Mickey Mouse's friend Goofy feeds the hens?
Are there really ten?

11, 12
Eleven, twelve,
Dig and delve.

Snow White's seven dwarfs and some friends
work in their mine.

13, 14
Thirteen, fourteen,
Mice a-courting.

When Cinderella lived with her stepmother, her best friends
were the mice who lived in the attic of her home.

15, 16
Fifteen, sixteen,
Here's a lively circus scene!

Dumbo, the baby elephant, is flying over the circus.
Everyone is watching.

How many circus animals and performers can you count?

17, 18

Seventeen, eighteen,
Everyone's waiting.

Mickey Mouse is treating all his friends
to a ride on the merry-go-round.

19, 20
Nineteen, twenty,
Mickey's purse is empty.

Mickey has so many friends! How many can you count?
Buying the tickets takes all of Mickey's money.
But what a good time for everyone!

MARCH BY TWOS

Tomorrow is Parade Day.
Everyone will march
in the wonderful big parade.

Mickey Mouse is the leader
of the big parade.

"We will line up by twos,"
Mickey says.

Any number that can
be divided by two—
or line up by twos—
is an even number.

Any number that cannot
divide into sets of two
is an odd number.

All the Disneyland folk have come to march in the parade. Can you tell which groups divide evenly into twos? Which ones have odd numbers?

Here are Minnie Mouse
and young Morty and Ferdie Mouse.
Can they march by twos?

Here are Donald and Daisy Duck.
Do they make a pair?

How about the three little pigs?

Here are Goofy and Pluto,
and Horace Horsecollar and
Clarabelle Cow.
Can they march by twos?

Cinderella and Prince Charming
have come in a golden coach.
Cinderella's Fairy Godmother
is with them.
Can they march in pairs?

Pinocchio and his father Geppetto
are here, along with Jiminy Cricket
and the Blue Fairy.
Do they make an even number?

Princess Briar Rose,
the Sleeping Beauty, is here
with her prince, too.
Can you see the three good fairies?
Nearby is the bad fairy.
Do they need her to even up
their number?
How many are they, all together?
Is that an even or odd number?

Mowgli and the jungle animals
have come to march, too.
There are Baloo, the bear, and
Bagheera, the black panther.
Ka, the sly python, is slithering along.
And here comes the great tiger, Shere Khan.
Can they line up by twos?

Nearby, the jungle elephants
are lining up to parade.
How many are they?
Is that an even number?

How many Monkeys have come?
Can they march by twos?

Peter Pan plans to fly along
with Wendy and John and Michael
and the Lost Boys of Never Land.
Dumbo and Timothy Mouse
will fly along with them.
Can they fly in pairs?
Will it help if Tinker Bell
(who is sulking on the toadstool)
joins them?
How many are they all together?

Marching with Snow White
and her prince are the seven dwarfs.
Do they make an even number?

Don't worry! Mickey Mouse will get everyone in order. He knows that some people don't mind marching alone.

69

TO MARKET, TO MARKET

Minnie Mouse was on her way
to market, to market
one fine day.
First came Mickey without a care,
on his ____-wheeled unicycle there.

Then as she continued her hike,
she met Donald Duck
on his ____-wheeled bike.

Along came Goofy,
cool as an icicle,
riding no-handed
on his ____-wheeled tricycle.

Next Minnie saw,
not very far,
Horace Horsecollar
in his ____-wheeled car.

Now, how many wheels were on their way to market, to market that fine day?

Here's another question that's more fair. How many wheels had Minnie **met**, going there?

Count them once and count again. Count all the way up to 10!